Fundamentals
of Natural Gas

Fundamentals of Natural Gas

Ed Etefia, P.E.

To order additional copies of this book, contact:
Xlibris
844-714-8691
www.Xlibris.com
Orders@Xlibris.com
788634

CONTENTS

Preface

This book is an updated version of my self-authored course for engineers, architects, scientists and technicians in the online continuing education RedVector Program that was contracted to the University of Tennessee a few years ago.

The intent is to inform the reader of the basics of natural gas as a desirable fossil fuel and its renewable derivatives, the mechanics of its production, and advances in exploration and production, with due consideration for the means of reducing emissions and adverse impacts on the environment.

I would like to express my appreciation to the Xlibris Publishing staff that assisted me so diligently from manuscript compilation to publication and marketing of this book.

My utmost gratitude also goes to my wife and family for their endurance and encouragement throughout this effort, and to the many readers who purchase this book to enhance their knowledge of natural gas as a useful alternative fossil fuel.

With Utmost Gratitude
Ed Etefia, P.E.

Introduction

This book is an overview of the rudimentary elements of natural gas. It informs the reader of the definition of natural gas, its uses and environmental considerations as a more desirable economical fossil fuel.

It offers a glimpse of its exploration and production technology options, such as 3-D and 4-D Seismic Imaging, Co2-Sand Fracturing, Coiled Tubing, and Hydraulic Fracturing to name a few, and provides valuable information on Liquefied Natural Gas (LNG) production and uses, Renewable Natural Gas production and uses, and the dependability and efficiency of Natural Gas Fuel Cells for clean electricity generation.

It also further informs on the environmental benefits of natural gas, while advising on practical strategies for mitigating adverse environmental impacts from the use of this important fuel.

I. Natural Gas: The Basics

What is natural gas?

Natural gas is a combustible, gaseous mixture of simple hydrocarbon compounds, usually found in deep underground reservoirs formed by porous rock.

It is an abundant, naturally occurring gas, which is found deep beneath the earth's surface, in large pockets that are located inside porous rock.

It is called a fossil fuel because scientists believe it to have been created by the gradual decomposition of ancient organic fossil matter, such as plants and tiny sea animals.

Layers of this organic matter built up over time until the pressure and heat from the earth "cooked" this mixture into natural gas.

Natural gas is composed almost entirely of methane, but does contain small amounts of other gases, including ethane, propane, butane and pentane. Methane is composed of a molecule of one carbon atom and four hydrogen atoms.

Natural gas is used extensively in residential, commercial and industrial applications. It is the dominant energy used for home heating with slightly more than one half of American homes using gas. Natural gas is odorless and colorless and produces very few emissions. It is considered the cleanest fossil fuel because of its clean-burning qualities.

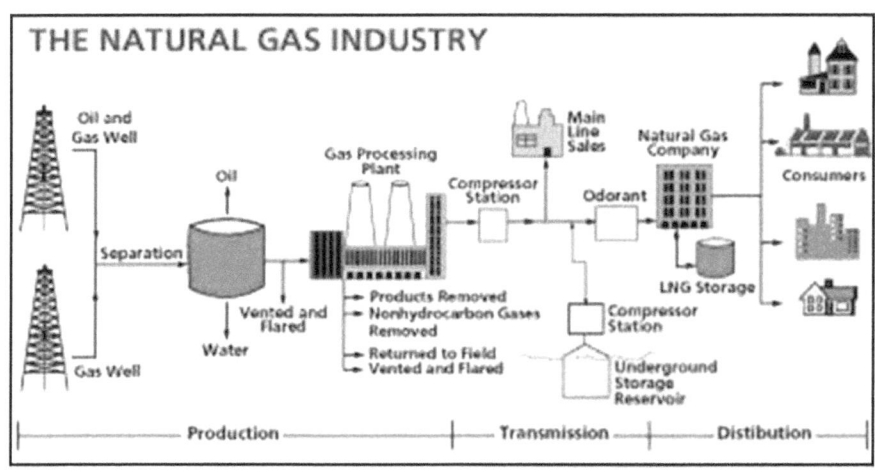

Source: U.S. Dept. of Energy, May 2015.

What are the uses of natural gas?

The natural gas delivery system brings natural gas to about 70 million homes and businesses in all 50 states.

Natural gas has become the most popular energy used for home heating: about 64 million American homes use natural gas because of its comfort, ease of use and efficiency.

Because of its environmental advantages due to low emissions, the use of natural gas is also rapidly increasing in electric power generation and cooling and as a transportation fuel.

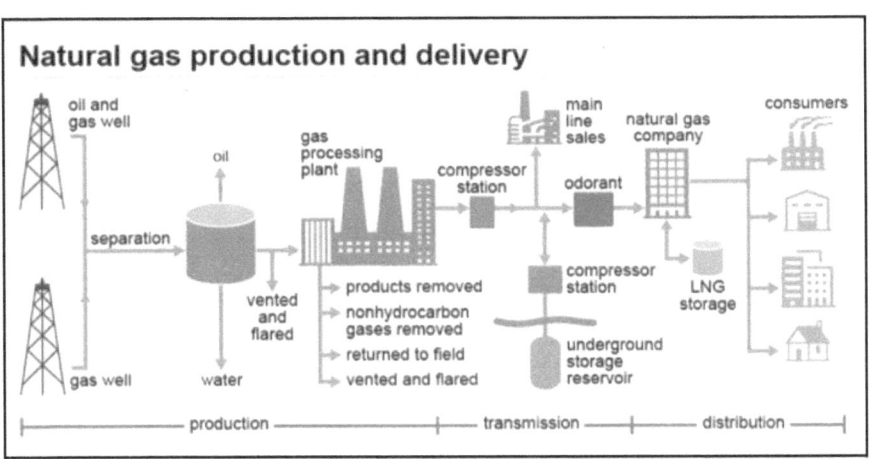

Source: U.S. Dept of Energy; Dec. 2016.

How safe is natural gas?

According to the U.S. Department of Transportation, the natural gas delivery system has the best safety record of any energy delivery system.

Improvements in technology and materials have contributed to a steady decline in natural gas pipeline-related incidents.

In fact, in recent years, the number of incidents on natural gas distribution pipelines decreased by more than 25 percent—yet the amount of natural gas traveling through the delivery system increased by 30 percent, and an additional 650,000 miles of pipeline were added to the system.

It is the dominant energy used for home heating with slightly more than one half of American homes using gas. Overall, more than 66 million homes use natural gas. The use of natural gas is also rapidly increasing in electric power generation and cooling.

Natural gas is the cleanest burning fossil fuel, producing primarily carbon dioxide, water vapor and small amounts of nitrogen oxides. Other fossil fuels are coal and oil, which together with natural gas, account for about 88 percent of U.S. energy consumption.

The prevailing scientific theory is that natural gas was formed millions of years ago when plants and tiny sea animals were buried by sand and rock.

Layers of mud, sand, rock and plant and animal matter continued to build up until the pressure and heat from the earth turned them into petroleum and natural gas.

The first use of gas energy in the United States occurred in 1816, when gaslights illuminated the streets of Baltimore, Md. By 1900, natural gas had been discovered in 17 states. During the years following World War II, expansion of the extensive interstate pipeline network occurred, bringing natural gas service to customers all o0ver the country.

Currently, oil provides the largest share of U.S. energy consumption-- about 39 percent of the entire market. Natural gas provides about 23 percent, coal 27 percent, hydropower 3 percent and nuclear power 8

percent. However, more than one-half of the oil Americans use is imported; in contrast, 82 percent of the natural gas U.S. consumers use is produced domestically. The rest comes from Canada via pipeline (14 percent) and LNG (4 percent).

Natural gas, like other forms of heat energy, is measured in British thermal units or Btu. One Btu is equivalent to the heat needed to raise the temperature of one pound of water by one degree Fahrenheit at atmosphere pressure.

A cubic foot of natural gas has about 1,027 Btu. Natural gas is normally sold from the wellhead in the production field to purchasers in standard volume measurements of thousands of cubic feet (Mcf). However, consumer bills are usually measured in heat content or therms. One therm is a unit of heating equal to 100,000 Btu.

Natural gas is delivered to about 170 million American consumers through a 1.5 million-mile network of underground pipe. A total of 450,000 producing natural gas wells, 125 natural gas pipeline companies and more than 1,200 gas distribution companies provide gas service to all 50 states. The United States accounts for about 19 percent of the world's natural gas production each year.

Three segments of the natural gas industry are involved in delivering natural gas from the wellhead to the consumer. Production companies explore, drill and extract natural gas from the ground. Transmission companies operate the pipelines that link the gas fields to major consuming areas. Distribution companies are the local utilities that deliver natural gas to the customer.

About 33 percent of natural gas delivered to U.S. consumers is used in the industrial sector, providing energy for everything from mining minerals to processing food. Generating electricity consumes about 31 percent. Another 15 percent is used in the commercial market -- for heating and cooling office buildings, hospitals and schools, and for cooking in restaurants. Most of the remaining amount -- about 22 percent -- is used in the residential market, providing energy for home heating, hot water, and cooking.

Emissions

i. Greenhouse Gas Emissions

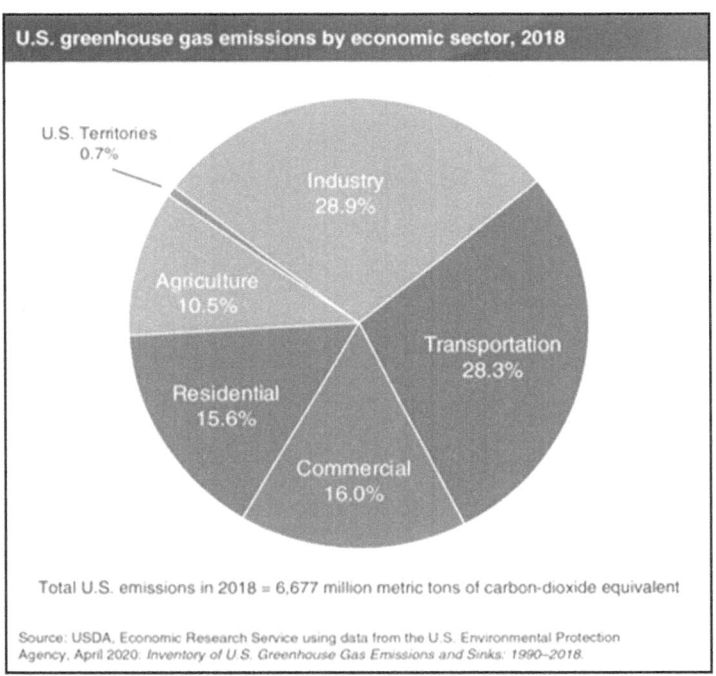

Gases that trap heat in the atmosphere are called greenhouse gases. The main greenhouse gases include:

Carbon dioxide (CO2): Carbon dioxide enters the atmosphere through burning fossil fuels (coal, natural gas and oil), solid waste, trees and wood products, and also as a result of certain chemical reactions (e.g., manufacture of cement).

Carbon dioxide is removed from the atmosphere (or "sequestered") when it is absorbed by plants as part of the biological carbon cycle.

Methane (CH4): Methane is emitted during the production and transport of coal, natural gas, and oil. Methane emissions also result from livestock and other agricultural practices and by the decay of organic waste in municipal solid waste landfills.

Nitrous oxide (N2O): Nitrous oxide is emitted during agricultural and industrial activities, as well as during combustion of fossil fuels and solid waste.

Fluorinated gases: Hydrofluorocarbons, perfluorocarbons, and sulfur hexafluoride are synthetic, powerful greenhouse gases that are emitted from a variety of industrial processes.

Fluorinated gases are sometimes used as substitutes for stratospheric ozone-depleting substances (e.g., chlorofluorocarbons, hydrochlorofluorocarbons, and ha lons).

These gases are typically emitted in smaller quantities, but because they are potent greenhouse gases, they are sometimes referred to as High Global Warming Potential gases ("High GWP gases").

Each gas's effect on climate change depends on three main factors:

How much of these gases are in the atmosphere?

Concentration, or abundance, is the amount of a particular gas in the air. Larger emissions of greenhouse gases lead to higher concentrations in the atmosphere.

Greenhouse gas concentrations are measured in parts per million, parts per billion, and even parts per trillion. One part per million is equivalent to one drop of water diluted into about 13 gallons of liquid (roughly the fuel tank of a compact car).

How long do they stay in the atmosphere?

Each of these gases can remain in the atmosphere for different amounts of time, ranging from a few years to thousands of years. All of these gases remain in the atmosphere long enough to become well mixed, meaning that the amount that is measured in the atmosphere is roughly the same all over the world, regardless of the source of the emissions.

How strongly do they impact global temperatures? Some gases are more effective than others at making the planet warmer and "thickening the Earth's blanket."

For each greenhouse gases, a Global Warming Potential (GWP) has been calculated to reflect how long it remains in the atmosphere, on average, and how strongly it absorbs energy. Gases with a higher GWP absorb more energy, per pound, than gases with a lower GWP, and thus contribute more to warming Earth.

Note: All emission estimates are from the ***Inventory of U.S. Greenhouse Gas Emissions***

ii. **Smog, Air Quality, and Acid Rain Air pollution,** which can lead to smog and acid rain, is the result of adding compounds or particles to the air that are harmful to human health or the environment. The most obvious danger of air pollution is that humans and other animals inhale pollutants and can become ill. In particular, air pollution leads to lung and respiratory illnesses —including asthma, bronchitis, and emphysema — and various cancers.

Air pollution causes smog

Smog originally described the specific combination of smoke and fog that discolored the air as a result of coal burning during the Industrial Revolution, but *smog* no longer means simply smoke and fog. Nowadays, the term *smog* refers to the complex combination of primary and secondary pollutants that turn the air a brown or yellow color.

Although smog isn't isolated to urban areas, it's more common around cities. One factor that intensifies smog in urban areas is the occurrence of a *thermal inversion* or *temperature inversion* in the atmosphere.

Generally speaking, the temperature of the air becomes gradually cooler as you move upward in the atmosphere; warm air near the surface moves upward, gradually cooling.

In the case of a temperature inversion, however, atmospheric circulation and geographic factors trap a layer of warm air between two layers of cooler air, inverting or flipping the usual pattern. This figure illustrates what this effect looks like.

Temperature inversions are most common in valleys where cool mountain air sweeps down into the valley at night, below the warm, polluted air surrounding the city. Los Angeles commonly experiences temperature inversions that trap a smog layer with warm air above and cool air below.

This inversion keeps the smog (and therefore the pollutants) close to the ground instead of allowing it to disperse into the atmosphere. Cities that experience such trapped smog may issue local air quality warnings so that people with asthma or other respiratory troubles know to stay indoors.

Air pollution causes acid rain.

Air pollution creates conditions in the atmosphere that change the pH of rainwater (and snow and other precipitation), making it more acidic. This ***acid rain*** is an environmental hazard because acidic rainwater damages whatever it falls on.

People first noticed the damage caused by acid rain more than 100 years ago. Folks in the mid-19th century observed that rainfall in heavily polluted cities (such as London during the Industrial Revolution) was dissolving marble and limestone statues. Not until the 1960s did scientists begin to carefully study the sequence of chemical reactions that results in acid rain.

Changing the pH of aquatic ecosystems may also have the side effect of allowing other contaminants (such as metals or toxins) to dissolve into the water and move more freely around the environment.

Normal atmospheric water has a pH of approximately 5.6 because of the natural formation of carbonic acid from atmospheric carbon dioxide gas. Pollutants such as nitrogen oxides and sulfur compounds in the atmosphere create atmospheric water particles that are more acidic than normal with pH values less than 5.

iii. Pollution from Industry and Electric Generation

The Most Polluting Industry

- Electric power generation produces more pollution than any other single industry in the United States.

- Burning fossil fuels such as coal or oil creates undesirable by-products that pollute when released into our environment, changing the earth's climate and harming ecosystems.

- Fossil fuels such as coal, oil and natural gas (most commonly used for electricity production) are known as non-renewable resources. Once burned to produce electricity, they are gone for ever.

According to 2000 figures, the U.S. electricity production industry is responsible for:

62.6 % of U.S. sulfur dioxide emissions that contribute to acid rain.

2 1.1% of U.S. nitrous oxide emissions that contribute to urban smog.

40% of U.S. carbon emissions that contribute to global climate change.

Water impacts, waste generation and land use disruption are among other major environmental issues associated with electricity.

By-products of electricity generation

- Nitrous oxide emissions: contributes to ground-level ozone, particulate matter pollution, haze pollution in national parks and wilderness areas, brown clouds in major western cities, the eutrophication of coastal waters, and acid deposition in sensitive ecosystems all over the country.

- Elevated ozone levels persisting throughout the country have also resulted in adverse health effects of smog and millions of dollars in agricultural damage.

By far, the worst ecosystem damage acid rain causes occurs in aquatic and wetland ecosystems. Acid rain creates acidic conditions in lakes, ponds, rivers, and wetlands, and the organisms in these ecosystems aren't adapted to survive in that kind of environment.

Acidification, or increasing the acidity (lowering the pH) of an aquatic ecosystem, can kill aquatic organisms such as fish and amphibians as well as interfere with their life cycles.

Natural gas is an extremely important source of energy for reducing pollution and maintaining a clean and healthy environment. In addition to being a domestically abundant and secure source of energy, the use of natural gas also offers a number of environmental benefits over other sources of energy, particularly other fossil fuels.

Emissions from the Combustion of Natural Gas

Natural gas is the cleanest of all the fossil fuels, as evidenced in the Environmental Protection Agency's data comparisons in the chart below, which is still current as of 2010. Composed primarily of methane, the main products of the combustion of natural gas are carbon dioxide and water vapor, the same compounds we exhale when we breathe. Coal and oil are composed of much more complex molecules, with a higher carbon ratio and higher nitrogen and sulfur contents. This means that when combusted, coal and oil release higher levels of harmful emissions, including a higher ratio of carbon emissions, nitrogen oxides (NOx), and sulfur dioxide (SO2). Coal and fuel oil also release ash particles into the environment, substances that do not burn but instead are carried into the atmosphere and contribute to pollution. The combustion of natural gas, on the other hand, releases very small amounts of sulfur dioxide and nitrogen oxides, virtually no ash or particulate matter, and lower levels of carbon dioxide, carbon monoxide, and other reactive hydrocarbons.

Fossil Fuel Emission Levels

- Pounds per Billion Btu of **Energy** Input

Pollutant	Natural Gas	Oil	Coal
Carbon Dioxide	117,000	164,000	208,000
Carbon Monoxide	40	33	208
Nitrogen Oxides	92	448	457
Sulfur Dioxide	1	1,122	2,591
Particulates	7	84	2,744
Mercury	0.000	0.007	0.016

Source: EIA - Natural Gas Issues and Trends 1998

Natural gas, as the cleanest of the fossil fuels, can be used in many ways to help reduce the emissions of pollutants into the atmosphere. Burning natural gas in the place of other fossil fuels emits fewer harmful pollutants, and an increased reliance on natural gas can potentially reduce the emission of many of these most harmful pollutants.

Pollutants emitted in the United States, particularly from the combustion of fossil fuels, have led to the development of many pressing environmental problems.

Natural gas, emitting fewer harmful chemicals into the atmosphere than other fossil fuels, can help to mitigate some of these environmental issues. These issues include:

Greenhouse Gas Emissions

Gases that trap heat in the atmosphere are called greenhouse gases. They include:

Carbon dioxide (CO_2): Carbon dioxide enters the atmosphere through burning fossil fuels (coal, natural gas and oil), solid waste, trees and wood products, and also as a result of certain chemical reactions (e.g., manufacture of cement). Carbon dioxide is removed from the atmosphere (or "sequestered") when it is absorbed by plants as part of the biological carbon cycle.

Methane (CH_4): Methane is emitted during the production and transport of coal, practices and by the decay of organic waste in municipal solid waste landfills.

Nitrous oxide (N_2O): Nitrous oxide is emitted during agricultural and industrial activities, as well as during combustion of fossil fuels and solid waste.

Fluorinated gases: Hydrofluorocarbons, perfluorocarbons, and sulfur hexafluoride are synthetic, powerful greenhouse gases that are emitted from a variety of industrial processes. Fluorinated gases are sometimes used as substitutes for stratospheric ozone-depleting substances (e.g., chlorofluorocarbons, hydrochlorofluorocarbons, and halons). These gases are typically emitted in smaller quantities, but because they are potent greenhouse gases, they are sometimes referred to as High Global Warming Potential gases ("High GWP gases").

Each gas's effect on climate change depends on three main factors:

Greenhouse Gas Emissions

How much of these gases are in the atmosphere?

Concentration, or abundance, is the amount of a particular gas in the air. Larger emissions of greenhouse gases lead to higher concentrations in the atmosphere. Greenhouse gas concentrations are measured in parts per million, parts per billion, and even parts per trillion. One part per million is equivalent to one drop of water diluted into about 13 gallons of liquid (roughly the fuel tank of a compact car).

How long do they stay in the atmosphere?

Each of these gases can remain in the atmosphere for different amounts of time, ranging from a few years to thousands of years. All of these gases remain in the atmosphere long enough to become well mixed, meaning that the amount that is measured in the atmosphere is roughly the same all over the world, regardless of the source of the emissions.

Some gases are more effective than others at making the planet warmer and "thickening the Earth's blanket." For each greenhouse gas, a Global Warming Potential (GWP) has been calculated to reflect how long it remains in the atmosphere, on average, and how strongly it absorbs energy. Gases with a higher GWP absorb more energy, per pound, than gases with a lower GWP, and thus contribute more to warming Earth.

Greenhouse Gas Emissions

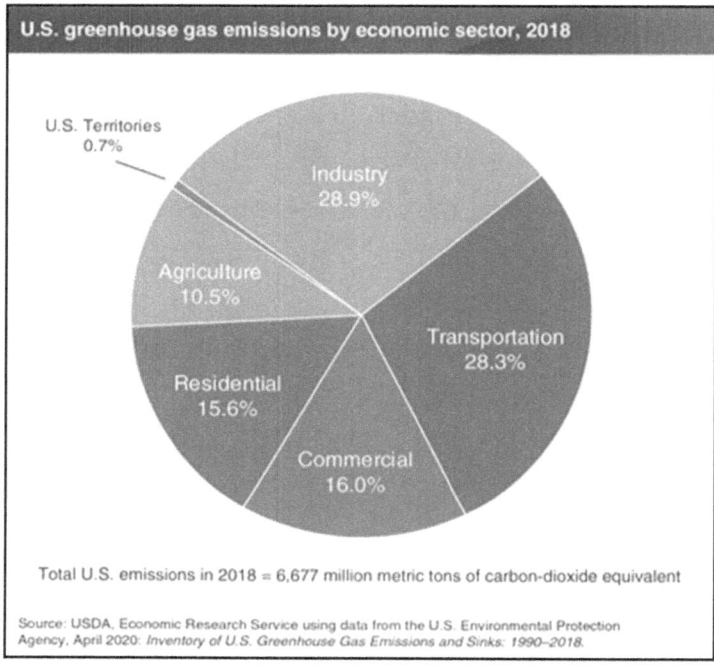

2018 US GREEN HOUSE GAS EMISSIONS

Global warming, or the 'greenhouse effect' is an environmental issue that deals with the potential for global climate change due to increased levels of atmospheric 'greenhouse gases'. There are certain gases in our atmosphere that serve to regulate the amount of heat that is kept close to the earth's surface.

Scientists theorize that an increase in these greenhouse gases will translate into increased temperatures around the globe, which would result in many disastrous environmental effects. In fact, the Intergovernmental Panel on Climate Change (IPCC) predicts in its *Fourth Assessment Report* released in 2007 that during the 21st century, global average temperatures are expected to rise by between 2.0 and 11.5 degrees Fahrenheit. A Fifth Assessment Report is expected to be released by the IPCC between 2013 and 2015.

Power Plants Contribute to the Emission of Greenhouse Gases
Source: API

The principal greenhouse gases include water vapor, carbon dioxide, methane, nitrogen oxides, and some engineered chemicals such as cholorofluorocarbons.

While most of these gases occur in the atmosphere naturally, levels have been increasing due to the widespread burning of fossil fuels by growing human populations. The reduction of greenhouse gas emissions has become a primary focus of environmental programs in countries around the world.

One of the principal greenhouse gases is carbon dioxide.

Although carbon dioxide does not trap heat as effectively as other greenhouse gases (making it a less potent greenhouse gas), the sheer volume of carbon dioxide emissions into the atmosphere is very high, particularly from the burning of fossil fuels. In fact, according to the Energy Information Administration in its December 2009 report *'Emissions of Greenhouse Gases'* in the United States, 81.3 percent of greenhouse gas emissions in the United States in 2008 came from energy-related carbon dioxide.

Because carbon dioxide makes up such a high proportion of U.S. greenhouse gas emissions, reducing carbon dioxide emissions can play a pivotal role in combating the greenhouse effect and global warming. The combustion of natural gas emits almost 30 percent less carbon dioxide than oil, and just under 45 percent less carbon dioxide than coal.

One issue that has arisen with respect to natural gas and the greenhouse effect is the fact that methane, the principle component of natural gas, is itself a potent greenhouse gas. Methane has an ability to trap heat almost 21 times more effectively than carbon dioxide.

According to the Energy Information Administration, although methane emissions account for only 1.1 percent of total U.S. greenhouse gas emissions, they account for 8.5 percent of the greenhouse gas emissions based on global warming potential. Sources of methane emissions in the U.S. include the waste management and operations industry, the agricultural industry, as well as leaks and emissions from the oil and gas industry itself.

A major study performed by the Environmental Protection Agency (EPA) and the Gas Research Institute (GRI), now Gas Technology Institute, in 1997 sought to discover whether the reduction in carbon dioxide emissions from increased natural gas use would be offset by a possible increased level of methane emissions. The study concluded that the reduction in emissions from increased natural gas use strongly outweighs the detrimental effects of increased methane emissions. More recently in 2011, researchers at the Carnegie Mellon University released "Life cycle greenhouse gas emissions of Marcellus shale gas", a report comparing greenhouse gas emissions from the Marcellus Shale region with emissions from coal used for electricity generation. The authors found that wells in the Marcellus region emit 20 percent to 50 percent less greenhouse gases than coal used to produce electricity.

In 1993, the natural gas industry joined with EPA in launching the Natural Gas STAR Program to reduce methane emissions. The STAR program has chronicled dramatic reductions to methane emissions, since that time:

EPA STAR data shows a reduction in methane emissions each year for the last 16 years More than 904 Billion cubic feet (Bcf) of methane emissions were eliminated through the STAR program 1993-2009; and In 2009 alone, the program reduced methane emissions by 86 Bcf.

Thus the increased use of natural gas in the place of other, dirtier fossil fuels can serve to lessen the emission of greenhouse gases in the United States.

Smog, Air Quality and Acid Rain

Smog and poor air quality is a pressing environmental problem, particularly for large metropolitan cities.

Smog, the primary constituent of which is ground level ozone, is formed by a chemical reaction of carbon monoxide, nitrogen oxides, volatile organic compounds, and heat from sunlight. As well as creating that familiar smoggy haze commonly found surrounding large cities, particularly in the summer time, smog and ground level ozone can contribute to respiratory problems ranging from temporary discomfort to long-lasting, permanent lung damage.

Pollutants contributing to smog come from a variety of sources, including vehicle emissions, smokestack emissions, paints, and solvents. Because the reaction to create smog requires heat, smog problems are the worst in the summertime.

Source: Environmental Protection Agency

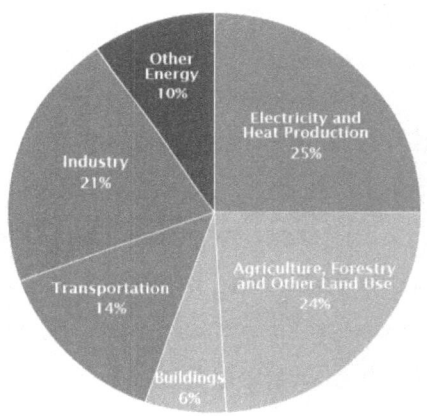

EMISSIONS BY INDUSTRY

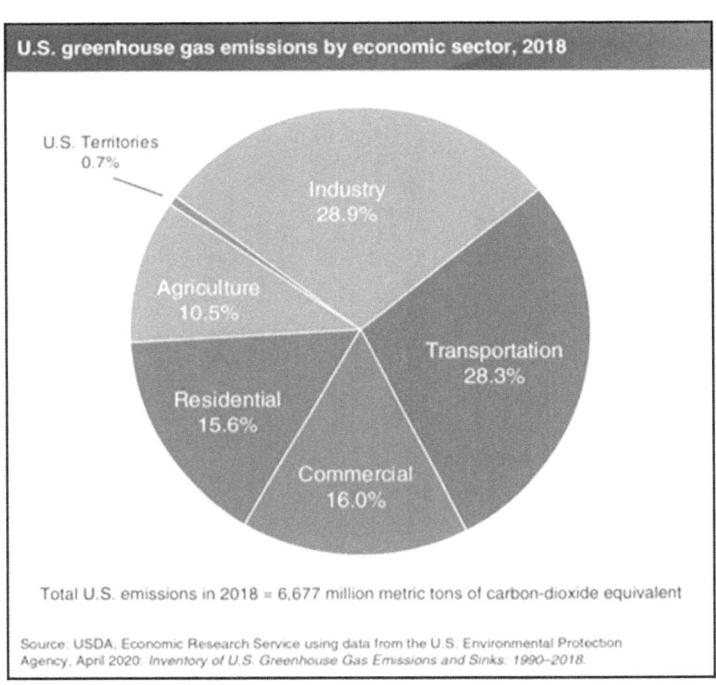

U.S. greenhouse gas emissions by economic sector, 2018

U.S. Territories
0.7%

Industry
28.9%

Agriculture
10.5%

Transportation
28.3%

Residential
15.6%

Commercial
16.0%

Total U.S. emissions in 2018 = 6,677 million metric tons of carbon-dioxide equivalent

Source: USDA, Economic Research Service using data from the U.S. Environmental Protection
Agency, April 2020: *Inventory of U.S. Greenhouse Gas Emissions and Sinks: 1990–2018.*

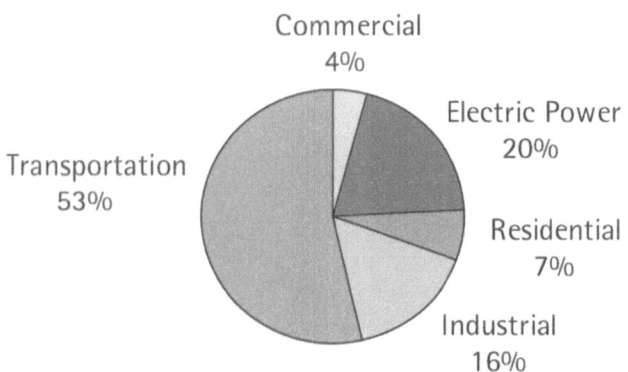

CO$_2$ emissions from fossil fuels in Oregon

Total: 40.4 million metric tons

Source: Environmental Protection Agency

Smog, Air Quality and Acid Rain

The use of natural gas does not contribute significantly to smog formation, as it emits low levels of nitrogen oxides, and virtually no particulate matter. For this reason, it can be used to help combat smog formation in those areas where ground level air quality is poor. The main sources of nitrogen oxides are electric utilities, motor vehicles, and industrial plants. Increased natural gas use in the electric generation sector, a shift to cleaner natural gas vehicles, or increased industrial natural gas use, could all serve to combat smog production, especially in urban centers where it is needed the most.

Particularly in the summertime, when natural gas demand is lowest and smog problems are the greatest, industrial plants and electric generators could use natural gas to fuel their operations instead of other, more polluting fossil fuels. This would effectively reduce the emissions of smog causing chemicals, and result in clearer, healthier air around urban centers.

Acid rain is another environmental problem that affects much of the Eastern United States, damaging crops, forests, wildlife populations, and causing respiratory and other illnesses in humans. Acid rain is formed when sulfur dioxide and nitrogen oxides react with water vapor and other chemicals in the presence of sunlight to form various acidic compounds in the air. The principle source of acid rain-causing pollutants, sulfur dioxide and nitrogen oxides, are coal fired power plants. Since natural gas emits virtually no sulfur dioxide, and up to

80 percent less nitrogen oxides than the combustion of coal, increased use of natural gas could provide for fewer acid rain causing emissions.

Particulate emissions also cause the degradation of air quality in the United States. These particulates can include soot, ash, metals, and other airborne particles. A study by the **Union of Concerned Scientists** in 1998, entitled 'Cars and Trucks and Air Pollution', showed that the risk of premature death for residents in areas with high airborne particulate matter was 26 percent greater than for those in areas with low particulate levels. Natural gas emits virtually no particulates into the atmosphere: in fact, emissions of particulates from natural gas combustion are 90 percent lower than from the combustion of oil, and 99 percent lower than burning coal.

Thus increased natural gas use in place of other dirtier hydrocarbons can help to reduce particulate emissions in the U.S. Current consequences stemming from global warming raised by the Union of Concerned Scientists can be found on their site.

Industrial and Electric Generation Emissions

Pollutant emissions from the industrial sector and electric utilities contribute greatly to environmental problems in the United States.

The use of natural gas to power both industrial boilers and processes and the generation of electricity can significantly improve the emissions profiles for these two sectors.

Industrial and Electric Generation Emissions

Natural gas is becoming an increasingly important fuel in the generation of electricity. As well as providing an efficient, competitively priced fuel for the generation of electricity, the increased use of natural gas allows for the improvement in the emissions profile of the electric generation industry. According to the National Environmental Trust (NET), now Pew Charitable Trusts (PEW), in their 2002 publication entitled 'Cleaning up Air Pollution from America's Power Plants', power plants in the U.S. account for 67 percent of sulfur dioxide emissions, 40 percent of carbon dioxide emissions, 25 percent of nitrogen oxide emissions, and 34 percent of mercury emissions. Coal fired power plants are the greatest contributors to these types of emissions. In fact, according to World Watch Report 184, natural gas combined cycle power plants emit half as much carbon dioxide as modern super critical coal plants.

Natural gas-fired electric generation and natural gas-powered industrial applications offer a variety of environmental benefits and environmentally friendly uses, including:

Fewer Emissions - Combustion of natural gas, used in the generation of electricity, industrial boilers, and other applications, emits lower levels of NOx, CO_2, and particulate emissions, and virtually no SO_2 and mercury emissions. Natural gas can be used in place of, or in addition to, other fossil fuels, including coal, oil, or petroleum coke, which emit significantly higher levels of these pollutants.

Industrial and Electric Generation Emissions

- **Reduced Sludge** – Coal-fired power plants and industrial boilers that use scrubbers to reduce SO2 emissions levels generate thousands of tons of harmful sludge. Combustion of natural gas emits extremely low levels of SO2, eliminating the need for scrubbers, and reducing the amounts of sludge associated with power plants and industrial processes.

- **Reburning** - This process involves injecting natural gas into coal or oil fired boilers. The addition of natural gas to the fuel mix can result in NOx emission reductions of 50 to 70 percent, and SO2 emission reductions of 20 to 25 percent.

Renewable Natural Gas (RNG) As A Viable Substitute For Fossil Natural Gas

RNG is anaerobically-generated biogas and renewable energy resource that is used to reduce emissions and provide environmental benefits. It has been enhanced (refined) as a substitute for fossil natural gas. Biogas is derived from organic sources such as municipal landfill solid waste , anaerobically digested sewage sludge, anaerobically digested yard and crop wastes, and food and food processing wastes and manure. Organically (from plants and animals) derived biogas typically contains 45 to 65% of methane (CH_4), depending on the source of the biogas, and must be refined through a series of stages to be converted into RNG.

The process of conversion of biogas to RNG involves an initial treatment to remove moisture, Carbon Dioxide (CO_2), particulates, trace-level contaminants (including siloxanes (a functional group in organosilicon chemistry with the Si-O-Si linkage), Volatile Organic Compounds (VOCs) and hydrogen sulfide (H_2S)), and reducing the Oxygen (O_2),

Nitrogen (N2) content. Upon purification, the RNG contains 90% or greater methane (CH4).   For RNG to be injected to a natural gas pipeline, its CH4 content must be between 96 and 98%.

RNG has numerous potential uses as a substitute for fossil natural gas. It can be used as a vehicle fuel, to produce electricity, in thermal applications, or as a bio-product feedstock. It can be directly injected into natural gas transmission of distribution pipelines, or used locally (at or near the production site).

The various uses of renewal natural gas and its delivery options are shown in the following images from the EPA.

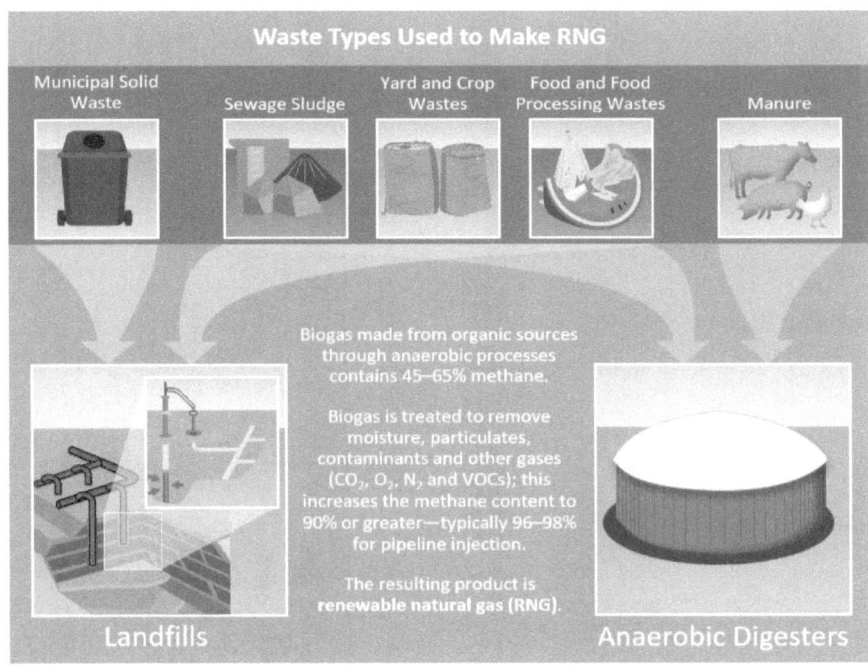

Figure 1. Organic Waste Types Used to Make RNG

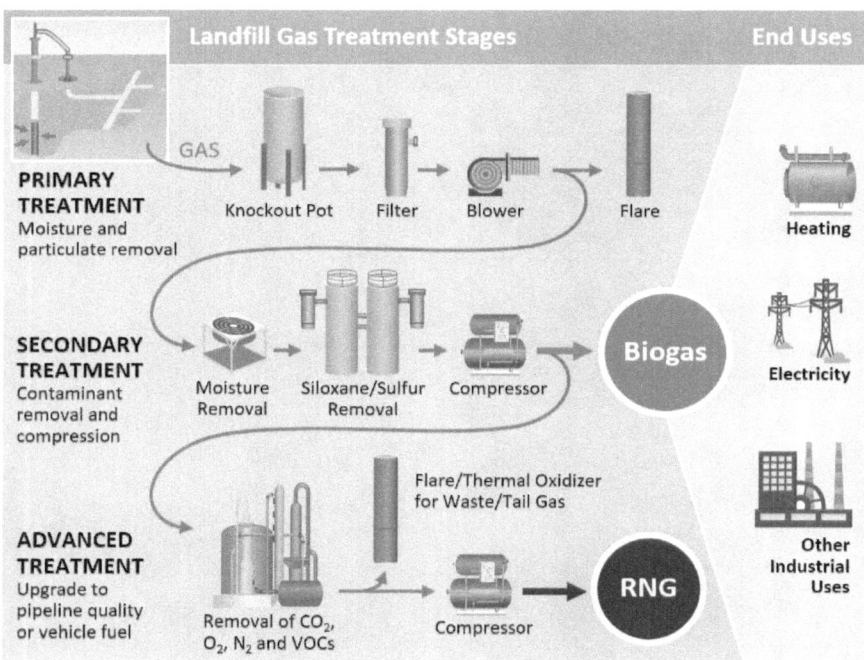

Figure 2. LFG Treatment Stages and Biogas End Uses

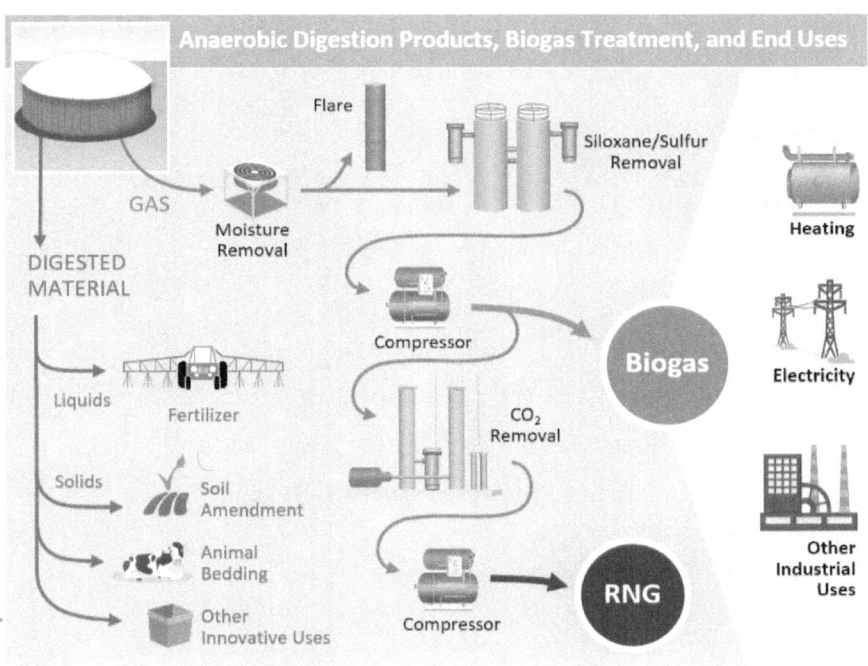

Figure 3. AD Products, Biogas Treatment and End Uses

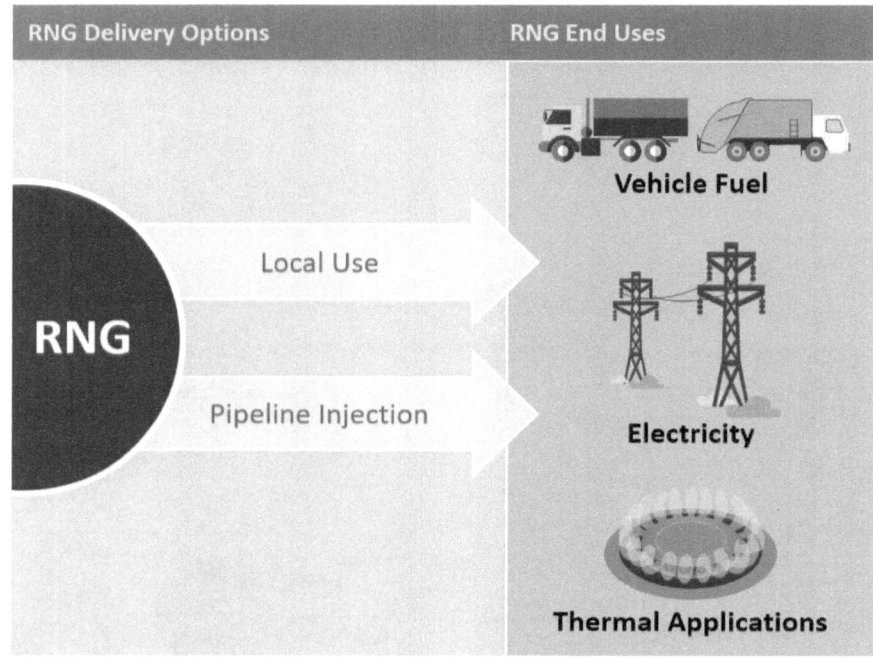

Figure 4. RNG Delivery Options
and Typical RNG End Uses

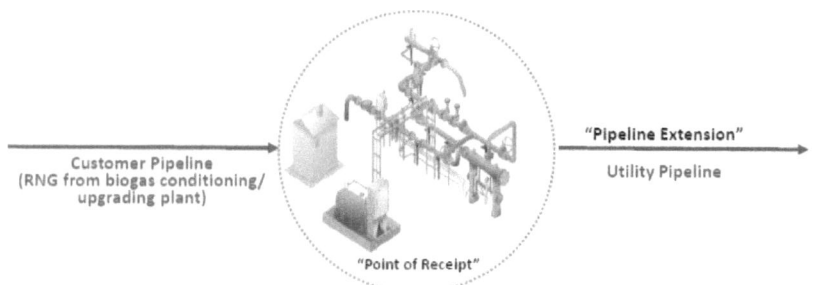

"Interconnection" = "Point of Receipt" + "Pipeline Extension"

Figure 5. Components of a Pipeline Interconnection

How A Combined Cycle Generation System Works

Combined cycle power plants feature gas and steam turbines. The gas turbine generates electricity using natural gas fuel, while the steam turbine generates electricity using waste heat from the gas turbine. The process is extremely efficient since exhaust heat is re-used that would otherwise be lost through the exhaust stack. A gas turbine compresses air and mixes it with fuel. The fuel is burned and the resultant hot air-fuel mixture is expanded through turbine blades, making them spin about a shaft. The spinning turbine drives a generator that converts the spinning energy into electricity.

Fuel is burned in a combustor. The resulting energy in the gas turbine turns the generator drive shaft Exhaust heat from the gas turbine is sent to a heat recovery steam generator (HRSG) The HRSG creates steam using the gas turbine exhaust heat and delivers it to the steam turbine The steam turbine delivers additional energy to the generator drive shaft The generator converts the energy into electricity.

Cogeneration - The production and use of both heat and electricity can increase the energy efficiency of electric generation systems and industrial boilers, which translates to the combustion of less fuel and the emission of fewer pollutants. Natural gas is the preferred choice for new cogeneration applications.

Combined Cycle Generation

Combined Cycle Generation generation units generate electricity and capture normally wasted heat energy, using it to generate more electricity. Like cogeneration applications, this increases energy efficiency, uses less fuel, and thus produces fewer emissions. Natural gas-fired combined-cycle generation units can be up to 60 percent energy efficient, whereas coal and oil generation units are typically only 30 to 35 percent efficient.

Fuel Cells - Natural gas fuel cell technologies are in development for the generation of electricity. Fuel cells are sophisticated devices that use hydrogen to generate electricity, much like a battery. No emissions are involved in the generation of electricity from fuel cells, and natural gas, being a hydrogen rich source of fuel, can be used. Although still under development, widespread use of fuel cells could in the future significantly reduce the emissions associated with the generation of electricity.

Essentially, electric generation and industrial applications that require energy, particularly for heating, use the combustion of fossil fuels for that energy. Because of its clean burning nature, the use of natural gas wherever possible, either in conjunction with other fossil fuels, or instead of them, can help to reduce the emission of harmful pollutants.

According to the Congressional Research Service's 2010 report: *"Displacing Coal with Generation from Existing Natural-Gas Fired Power Plants,"* if natural-gas combined cycle plants utilization were to be doubled from 42 percent capacity factor to 85 percent, then the amount of power generated would displace 19 percent of the CO_2 emissions attributed to coal-fired electricity generation.

Pollution from the Transportation Sector - Natural Gas Vehicles

The transportation sector (particularly cars, trucks, and buses) is one of the greatest contributors to air pollution in the United States. Emissions from vehicles contribute to smog, low visibility, and various greenhouse gas emissions. According to the Department of Energy (DOE), about half of all air pollution and more than 80 percent of air pollution in cities are produced by cars and trucks in the United States. Currently, automobile manufacturers are under pressure to produce more environmentally friendly vehicles.

Natural gas can be used in the transportation sector to cut down on these high levels of pollution from gasoline and diesel powered cars, trucks, and buses. According to the EPA, compared to traditional vehicles, vehicles operating on compressed natural gas have reductions in carbon monoxide emissions of 90 to 97 percent, and reductions in carbon dioxide emissions of 25 percent. Nitrogen oxide emissions can be reduced by 35 to 60 percent, and other non-methane hydrocarbon emissions could be reduced by as much as 50 to 75 percent. In addition, because of the relatively simple makeup of natural gas in comparison to traditional vehicle fuels, there are fewer toxic and carcinogenic emissions from natural gas vehicles, and virtually no particulate emissions.

Thus the environmentally friendly attributes of natural gas may be used in the transportation sector to reduce air pollution.

Natural gas vehicles represent a growing segment of the transportation sector. According to the Natural Gas Vehicle Coalition, the use of natural gas for vehicles doubled between 2003 and 2009. Over 100,000 natural gas vehicles are currently on US roads. A large portion of those vehicles are transit buses, which account for nearly 62 percent of all natural gas vehicles.

Natural gas is the cleanest of the fossil fuels, and thus its many applications can serve to decrease harmful pollution levels from all sectors, particularly when used together with or replacing other fossil fuels. The natural gas industry itself is also committed to ensuring that the process of producing natural gas is as environmentally-friendly as possible. The Natural Gas Vehicle Coalition has more information regarding natural gas-powered vehicles.

Advantages

The biggest advantage of NGVs is that they reduce environmentally harmful emissions. Natural-gas vehicles can achieve up to a 93 percent reduction in carbon monoxide emissions, 33 percent reduction in emissions of various oxides of nitrogen and a 50 percent reduction in reactive hydrocarbons when compared to gasoline vehicles. NGVs also rate higher in particulate matter 10 (PM10) emissions. PM10 particles transport and deposit toxic materials through the air. NGVs that operate in diesel applications can reduce PM10 emissions by a factor of 10. Natural-gas vehicles also offer these benefits:

NGVs are safer. The fuel storage tanks on an NGV are thicker and stronger than gasoline or diesel tanks. There has not been an NGV fuel-tank rupture in more than two years in the United States.

- Natural gas costs are lower than gasoline. On average, natural gas costs one-third less than gasoline at the pump.

- Natural gas is convenient and abundant. A well-established pipeline infrastructure exists in the United States to deliver natural gas to almost every urban area and most suburban areas. There are more than 1,300 NGV fueling stations in the United States, and more are being added every day.

- Natural gas prices have exhibited significant stability compared to oil prices. Historically, natural gas prices have exhibited significant price stability compared to the prices of petroleum-based fuels. This stability makes it easier to plan accurately for long-term costs. NGVs have lower maintenance costs. Because natural gas burns so cleanly, it results in less wear and tear on the engine and extends the time between tune-ups and oil changes.

Disadvantages

One of the biggest complaints about NGVs is that they aren't as roomy as gasoline cars. This is because NGVs have to give up precious cargo and trunk space to accommodate the fuel storage cylinders. Not only that, these cylinders can be expensive to design and build -- a contributing factor to the higher overall costs of a natural-gas vehicle compared to a gasoline-powered car.

Another drawback is the limited driving range of NGVs, which is typically about half that of a gasoline-powered vehicle. For example, Honda's natural gas Civic, the Civic GX, can go up to 220 miles withoutrefueling. A typical gasoline-powered Civic can go approximately 350 miles without refueling. If a dedicated NGV ran out of fuel on the road, it would have to be towed to the owner's home or to a local natural gas refueling station, which might be harder to find than a "regular" gas station.

Finally, it should be noted that natural gas, like gasoline, is a fossil fuel and cannot be considered a renewable resource. While natural gas reserves in the United States are still considerable, they are not inexhaustible. Some predict that there are enough natural gas reserves remaining to last another 67.1 years, assuming that the 2003 level of production continues.

Despite some of the advantages offered by NGVs, they are still relatively uncommon. According to the Natural Gas Coalition, there are currently 130,000 NGVs on the road in the United States today and more than 2.5 million worldwide. To put this into perspective, consider that there were 142.5 million registered vehicles in 2001 --which means gas-powered vehicles outnumber NGVs almost 1,100 to one in the United States. And yet more than 40 different manufacturers, including Ford, General Motors, Toyota and Volvo, currently produce NGVs.

II. Technological Applications

a. Advances in Exploration and Production

Technological innovation in the exploration and production sector has equipped the industry with the equipment and practices necessary to continually increase the production of natural gas to meet rising demand. These technologies serve to make the exploration and production of natural gas more efficient, safe, and environmentally friendly. Even as natural gas deposits are increasingly produced from "unconventional" formations such as shale rock, the exploration and production industry has not only kept up its production pace, but in fact has improved the general nature of its operations, contributing to an unprecedented 39 percent increase in the size of U.S. resources since 2006.

According to a Department of Energy Report, "***Environmental Benefits of Advanced Oil and Gas Exploration and Production Technology,***" released in 1999 and still one of the most in-depth analyses available as of 2010:

22,000 fewer wells are needed on an annual basis to develop the same amount of oil and gas reserves as were developed in 1985.

Had technology remained constant since 1985, it would take two wells to produce the same amount of oil and natural gas as one 1985 well. However, advances in technology mean that one well today can produce two times as much as a single 1985 well.

Drilling wastes have decreased by as much as 148 million barrels due to increased well productivity and fewer wells.

- The drilling footprint of well pads has decreased by as much as 70 percent due to advanced drilling technology, which is extremely useful for drilling in sensitive areas.

- By using modular drilling rigs and slimhole drilling, the size and weight of drilling rigs can be reduced by up to 75 percent over traditional drilling rigs, reducing their surface impact.

- Had technology, and thus drilling footprints, remained at 1985 levels, today's drilling footprints would take up an additional 17,000 acres of land.

- New exploration techniques and vibrational sources mean less reliance on explosives, reducing the impact of exploration on the environment.

New exploration techniques and vibrational sources mean less reliance on explosives, reducing the impact of exploration on the environment. New exploration techniques and vibrational sources mean less reliance on explosives, reducing the impact of exploration on the environment.

Some of the major recent technological innovations in the exploration and production sector include:

3-D and 4-D Seismic Imaging- The development of seismic imaging in three dimensions greatly changed the nature of natural gas exploration. This technology uses traditional seismic imaging techniques, combined with powerful computers and processors, to create a three-dimensional model of the subsurface layers. 4-D seismology expands on this, by adding time as a dimension, allowing exploration teams to observe how subsurface characteristics change over time.

Exploration teams can now identify natural gas prospects more easily, place wells more effectively, reduce the number of dry holes drilled, reduce drilling costs, and cut exploration time. This leads to both economic and environmental benefits.

3D seismic interpretation is a form of seismic interpretation which relies on the use of 3D surveys which provide visualizations of structures in three dimensions. People often use specialized software for this task, as 3D seismic interpretation requires a lot of math and the careful construction and interpretation of data. There are a number of applications for this process, including the examination of sites to determine whether or not they would make viable oilfields, exploration of the ocean floor, and general geological study. In seismic surveys, controlled explosions are generated and the reflections of these explosions are read to generate data about what is going on underground. With 3D seismic interpretation, this data is mapped on a three dimensional representation which allows people to explore the data in a number of different ways. Rather than visualizing a site in the form of a flat elevation map or cross section, 3D seismic interpretation allows people to manipulate the angle of view and to visualize a site as a whole. It can also provide information about the surrounding area which may not be readily apparent with other mapping techniques. Seismic interpretation can get very complex. Geologists are interested in the fundamental structure of the Earth, and they are also interested in the components of the sites they are studying. Different types of rock reflect explosions differently, and 3D seismic interpretation is designed to reveal not only the presence of underground formations, but what is in those formations, and where the transitions between different types of materials are occurring. Using this information, a geologist can play with scenarios. Modeling scenarios on a map allows geologists to explore the possible results of various activities. For example, a geologist may be concerned that oilfield exploration could cause the collapse of a delicate formation, potentially putting people or the environment in danger. They may also believe that formations present on site hold a limited amount of useful resources, making investment in the site potentially unprofitable.

Advanced 4-D Seismic Imaging

Time-lapse 3D, or 4D, seismic technology is the use of 3D seismic **surveys** acquired at different times in the productive life of a reservoir. The term 4D encompasses a broad workflow from feasibility and design, to acquisition and processing, to inversion and interpretation, and finally to integration with reservoir management, although a 4D project may span only a subset of the workflow.

4D with zero time-lapse, meaning no production-induced subsurface change, is a repeatability study. The time-lapse interval for production monitoring is reservoir- and process-dependent, and may vary considerably, from as short as two weeks to monitor the pressure change due to first oil, to many years in a large, Middle East carbonate reservoir. There is no general optimal time-lapse interval, so prediction of time-lapse changes at different times in the future is an important part of 4D feasibility studies, making use of flow simulation and rock physics models.

For optimal results, 4D requires time-lapse logs, special core analysis, VSPs, pressure, and other data sources such as multicomponent, geomechanical, gravimetric, and electromagnetic data, when available. These data serve to constrain the inversion and interpretation of 4D surveys, just as they do for 3D surveys.

CO2-Sand Fracturing - Fracturing techniques have been used since the 1970s to help increase the flow rate of natural gas and oil from underground formations. CO2-Sand fracturing involves using a mixture of sand propants and liquid CO_2 to fracture formations, creating and enlarging cracks through which oil and natural gas may flow more freely. The CO_2 then vaporizes, leaving only sand in the formation, holding the newly enlarged cracks open. Beca there are no other substances used in this type of fracturing, there are no 'leftovers' from the fracturing process that must be removed. This means that, while this type of fracturing effectively opens the formation and allows for increased recovery of oil and natural gas, it does not damage the deposit, generates no below ground wastes, and protects groundwater resources.

Coiled Tubing - Coiled tubing technologies replace the traditional rigid, jointed drill pipe with a long, flexible coiled pipe string. This greatly reduces the cost of drilling, as well as providing a smaller drilling footprint, requiring less drilling mud, faster rig set up, and reducing the time normally needed to make drill pipe connections. Coiled tubing can also be used in combination with slim hole drilling to provide very economic drilling conditions, and less impact on the environment.

Measurement While Drilling - (MWD) systems allow for the collection of data from the bottom of a well as it is being drilled. This allows engineers and drilling teams access to up-to-the-second information on the exact nature of the rock formations being encountered by the drill bit. This improves drilling efficiency and accuracy in the drilling process, allows better formation evaluation as the drill bit encounters the underground formation, and reduces the chance of formation damage and blowouts.

Slim hole Drilling – Slim hole drilling is exactly as it sounds; drilling a slimmer hole in the ground to get to natural gas and oil deposits. In order to be considered slim hole drilling, at least 90 percent of a well must be drilled with a drill bit less than six inches in diameter (whereas conventional wells typically use drill bits as large as 12.25 inches in diameter). Slim hole drilling can significantly improve the efficiency of drilling operations, as well as decrease its environmental impact. In fact, shorter drilling times and smaller drilling crews can translate into a 50 percent reduction in drilling costs, while reducing the drilling footprint by as much as 75 percent. Because of its low cost profile and reduced environmental impact, slim hole drilling provides a method of economically drilling exploratory wells in new areas, drilling deeper wells in existing fields, and providing an efficient means for extracting more natural gas and oil from un-depleted fields.

Offshore Drilling Technology -The offshore oil and natural gas production sector is sometimes compared to the aeronautics field and NASA due to achievements in deepwater drilling that have been facilitated by state of the art technology. New technology, including improved offshore drilling rigs, dynamic positioning devices and sophisticated navigation systems are allowing safe, efficient offshore drilling in waters more than 10,000 feet deep

Hydraulic Fracturing also called "Fracking," or "Frac'ing"- Used to free natural gas that is trapped in shale rock formations. A liquid mix that is 99 percent water and sand is injected into the rock at very high pressure, creating fractures within the rock that provide the natural gas a path to flow to the wellhead. The fracking fluid mix also helps to keep the formation more porous. Hydraulic fracturing is now widely used, with more than 90 percent of the natural gas wells in the United States having used it to boost production at some time.

The above technological advancements provide only a snapshot of the increasingly sophisticated technology being developed and put into practice in the exploration and production of natural gas and oil. New technologies and applications are being developed constantly, and serve to improve the economics of producing natural gas, allow for the production of deposits formerly considered too unconventional or uneconomic to develop, and ensure that the supply of natural gas keeps up with steadily increasing demand. Sufficient domestic natural gas resources exist to help fuel the U.S. for a significant period of time, and technology is playing a huge role in providing low-cost, environmentally sound methods of extracting these resources.

Two other technologies that are revolutionizing the natural gas industry include the increased use of liquefied natural gas, and natural gas fuel cells. These technologies are discussed below.

Liquefied Natural Gas

Cooling natural gas to about -260°F at normal pressure results in the condensation of the gas into liquid form, known as Liquefied Natural Gas (LNG). LNG can be very useful, particularly for the transportation of natural gas, since LNG takes up about one six hundredth the volume of gaseous natural gas. Advances in technology are reducing the costs associated with the liquification and regasification of LNG. Because it is easy to transport, LNG can serve to make economical stranded natural gas deposits from around the globe for which the construction of pipelines is uneconomical.

LNG, when vaporized to gaseous form, will only burn in concentrations of between 5 and 15 percent mixed with air. In addition, LNG, or any vapor associated with LNG, will not explode in an unconfined environment.

Thus, in the unlikely event of an LNG spill, the natural gas has little chance of igniting an explosion.

Liquification removes oxygen, carbon dioxide, sulfur, and water from the natural gas, resulting in LNG that is almost pure methane.

LNG is typically transported by specialized tanker with insulated walls, and is kept in liquid form by autorefrigeration, a process in which the LNG is kept at its boiling point, so that any heat additions are countered by the energy lost from LNG vapor that is vented out of storage and used to power the vessel The increased use of LNG is allowing for the production and marketing of natural gas deposits that were previously economically unrecoverable. Although it currently accounts for only about 1 percent of natural gas used in the United States, it is expected that LNG imports will provide a steady, dependable source of natural gas for U.S. consumption.

Natural Gas Fuel Cells

Fuel cells powered by natural gas are an extremely exciting and promising new technology for the clean and efficient generation of electricity. Fuel cells have the ability to generate electricity using electrochemical reactions as opposed to combustion of fossil fuels to generate electricity. Essentially, a fuel cell works by passing streams of fuel (usually hydrogen) and oxidants over electrodes that are separated by an electrolyte. This produces a chemical reaction that generates electricity without requiring the combustion of fuel, or the addition of heat as is common in the traditional generation of electricity. When pure hydrogen is used as fuel, and pure oxygen is used as the oxidant, the reaction that takes place within a fuel cell produces only water, heat, and electricity. In practice, fuel cells result in very low emission of harmful pollutants, and the generation of high-quality, reliable electricity. The use of natural gas-powered fuel cells has a number of benefits, including:

Clean Electricity - Fuel cells provide the cleanest method of producing electricity from fossil fuels. While a pure hydrogen, pure oxygen fuel cell produces only water, electricity, and heat, fuel cells in practice emit trace amounts of sulfur compounds and very low levels of carbon dioxide. However, the carbon dioxide produced by fuel cell use is concentrated and can be readily recaptured, as opposed to being emitted into the atmosphere.

Distributed Generation - Fuel cells can come in extremely compact sizes, allowing for their placement wherever electricity is needed. This includes residential, commercial, industrial, and even transportation settings.

- **Dependability** - Fuel cells are completely enclosed units, with no moving parts or complicated machinery. This translates into a dependable source of electricity, capable of operating for thousands of hours. In addition, they are very quiet and safe sources of electricity. Fuel cells also do not have electricity surges, meaning they can be used where a constant, dependable source of electricity is needed.

- **Efficiency** - Fuel cells convert the energy stored within fossil fuels into electricity much more efficiently than traditional generation of electricity using combustion. This means that less fuel is required to produce the same amount of electricity. The National Energy **Technology Laboratory** estimates that, used in combination with natural gas turbines, fuel cell generation facilities can be produced that will operate in the 1 to 20 Megawatt range at 70 percent efficiency, which is much higher than the efficiencies that can be reached by traditional generation methods within that output range.

Conclusion

It has been an honor and privilege to prepare this book on the rudiments of Natural Gas as an economical fossil fuel alternative.

It is intended to provide an overview of the natural gas industry in terms of the basic uses of natural gas and its variants, environmental considerations to enable adverse impact mitigation, advantages and disadvantages of its use, and related technological applications and advances, based on my extensive experience as a retired petroleum engineer in Texas, Oklahoma and New Mexico.

I therefore hope that this book would enlighten the readers and spur more interest in the industry.

Respectfully,
Ed

Bibliography

1. Natural Gas Production Engineering; Mohan Kelkar, June 2019.
2. Natural Gas Technologies R&D; U.S. Dept. of Energy, March 2021.
3. Advanced Natural Gas Engineering; Xiuli Wang & Michael Economides, June 2010.
4. Handbook of Natural Gas Analysis, 2nd Edition; James G. Speight, PhD, DSc; May 2019, GPP.
5. Natural Gas & Renewable Methane for Powertrains: Future Strategies for a Climate-Neutral Mobility; Helmut List, July 2020.
6. Handbook of Liquefied Natural Gas; Saeid Mokhatab, John Y. Mak, Jaleel V. Valappil & David A. Wood, GPP, Aug. 2021.
7. Oil & Gas 1 Student's Book; Lewis Lansford, Feb. 2017.
8. U.S. Oil & Natural Gas: Providing Energy Security and Supporting Our Quality of Life; U.S. Dept. of Energy, Office of Oil & Natural Gas, June 2021.
9. Natural Gas: A Promising Transition Fuel for Sustainable Energy Future; Energy Mgr Magazine, Oct.-Dec. 2012

www.ingramcontent.com/pod-product-compliance
Lightning Source LLC
Chambersburg PA
CBHW021503210526
45463CB00002B/873